LONDO
SEWE...

Paul Dobraszczyk

SHIRE PUBLICATIONS

Published in Great Britain in 2014 by Shire Publications Ltd, PO Box 883, Oxford, OX1 9PL, UK.

PO Box 3985, New York, NY 10185-3985, USA.

E-mail: shire@shirebooks.co.uk www.shirebooks.co.uk

A CIP catalogue record for this book is available from the British Library.

Shire Library no. 800. ISBN-13: 978 0 74781 431 3
PDF ebook ISBN: 978 0 74781 531 0
e-pub ISBN: 978 0 74781 530 3

Paul Dobraszczyk has asserted his right under the Copyright, Designs and Patents Act, 1988, to be identified as the author of this book.

Designed by Tony Trucott Designs, Sussex, UK and typeset in Perpetua and Gill Sans.

Printed in China through Worldprint Ltd.

14 15 16 17 18   10 9 8 7 6 5 4 3 2 1

COVER IMAGE
Cover design by Peter Ashley. Front cover: sewer maintenance men removing a trolley loaded with silt deposits from a London sewer, April 1950 (Getty Images). Back cover: a London manhole cover (photograph courtesy of Lynn Hall).

TITLE PAGE IMAGE
View inside a storm drain under Brockwell Park in Brixton.

CONTENTS PAGE IMAGE
Subterranean surveyors inspecting repairs to the Fleet sewer, as shown in the *Illustrated London News* in 1854.

ACKNOWLEDGEMENTS
Grateful thanks to all those who provided assistance, in particular Robin Winters and Matthew Wood at Thames Water plc, the staff at London Metropolitan Archives, all at Spire Books, and the following individuals: Jeff Brookes, Bryce Caller, Ben Campkin, Paul Davies, Lisa Dobraszczyk, Nick Driver, Paul Driver, Michael Dunmow, Stephen Halliday, Judi Loach, Benedict O'Looney, Chris Pierce, Bruno Rinvolucri, Mike Seabourne and Robert Thomas.

IMAGE ACKNOWLEDGEMENTS
Thanks are also due to the organisations who allowed copyright photographs to be reproduced: City of London, London Metropolitan Archives, pages 14, 16–22, 26, 49; Alamy, pages 40, 42 (bottom), and 46–7; George Rex (Flickr), page 43 (bottom); Thames Water plc, pages 5, 8, 23, 27, 43, 50. The remainder of the images are either from the author's own collection and photographs, or from Victorian periodicals that are not subject to copyright restrictions.

Shire Publications is supporting the Woodland Trust, the UK's leading woodland conservation charity, by funding the dedication of trees.

CONTENTS

INTRODUCTION

M OST OF US take for granted the existence of effective forms of waste disposal in cities like London; the sewers beneath the city are generally invisible spaces that we would prefer not to think about. Yet, we owe their very existence to a Victorian revolution no less momentous than that which brought about better-known underground spaces in London, such as the Tube. Indeed, in the mid-Victorian period, sewers were a topic of polite conversation, endlessly debated by both sanitary reformers and the general public. For it was the Victorians, and particularly one man, the engineer Sir Joseph Bazalgette (1819–91), who transformed London's sewers from a piecemeal and crumbling medieval system into the main drainage system of intercepting sewers that is still in place today.

Portrait of a young Joseph Bazalgette featured in the *Illustrated London News* in 1859, the year his main drainage system for London began to be constructed.

The London sewers form an intricate, complex structure. Mirroring Victor Hugo's famous description of the Paris sewers in his epic novel *Les Misérables* (1862), they can be thought of as being like a vast tree: the smallest twigs of that tree are the city's household drains; the larger branches the street sewers to which those smaller household drains connect; the largest branches and trunk the main drainage system of intercepting and outfall sewers, with the whole arrangement of twigs, branches and trunk representing the city's complete sewerage system. This book focuses on the development of the largest of these branches, that is, the main drainage system, which consists of very large sewers designed to intercept waste from existing street sewers and divert it from west to east across London, eventually discharging their contents into the River Thames outside the built-up area. For it was the construction of these sewers, masterminded by Bazalgette, which transformed current thinking on how to deal with human waste in a metropolis such as London.

The book will tell the story of this transformation, beginning with an overview of the sanitary crisis that precipitated it, when London was a city bedevilled with filth and the overpowering stench of human waste. Sanitary reformer Edwin Chadwick (1800–90) played a key role, drawing attention to London's lack of effective sanitation and the threat of disease (particularly cholera) and then remapping London, building new sewers and cleansing the old ones. Bazalgette's contribution was equally significant, resulting in a new system of sewers that he designed and whose construction he supervised, despite opposition and many setbacks. His magnificent pumping stations – cathedrals of sewage – formed a vital part of the new system as well as glorifying its achievements. Since their construction, London's sewers have inspired literature and film and are now being explored by the more adventurous. As well as being an extraordinary engineering achievement, it is clear that London's sewers are spaces that continue to stimulate the imagination.

Large-scale map of London's main sewer network as it was in 1934.

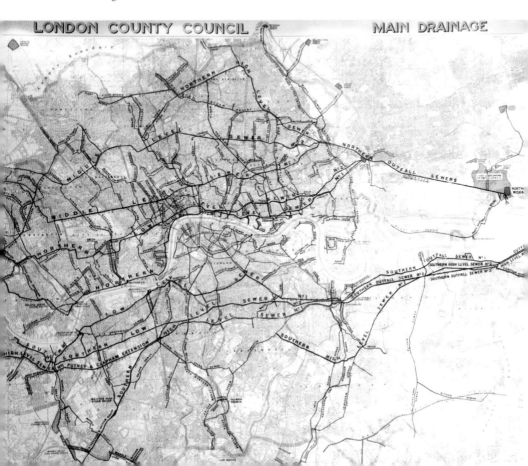

FARADAY GIVING HIS CARD TO FATHER THAMES

And we hope the Dirty Fellow will consult the learned Professor.

THE FILTHY CITY

L ONDON'S SEWERS have always been seen as somehow representative of the city as a disease-ridden and filthy environment. In the early seventeenth century, the poet Ben Jonson (1572–1637) made a journey along the River Fleet, which he famously committed to verse as 'On the Famous Voyage' (c. 1612), describing, in nauseating detail, the horrific sights and smells he encountered in what was already then effectively a pre-modern sewer. However, it was the exponential growth of London in the early nineteenth century that propelled the condition of its sewers into wider debates about the poor quality of the urban environment. With the population of the city increasing almost threefold in the first half of the nineteenth century (from just under 1 million to 3 million), the city's sanitation, or the lack of it, became a dominating concern. What had once been an effective and sustainable system of natural drainage in London was quickly overwhelmed by the sheer numbers of people now living (and disposing of their waste) in the city.

London's many rivers – tributaries of the Thames such as the Fleet, Westbourne and Tyburn – had, until the beginning of the nineteenth century, provided a ready means of draining rainwater within the built-up area of the city. The enormous expansion of London in the early nineteenth century led to these rivers being systematically built over and, coupled with the increasing (if illegal) practice of discharging human and other waste into watercourses, put an ever-increasing strain on this existing system, which had been in place for centuries.

Until the mid-nineteenth century, the normal method of disposing of human waste was to empty it into pits, known as cesspools, which were usually located close to dwellings and were often shared by several householders. These cesspools were periodically emptied and cleaned by so-called 'nightmen', that is, shady workers who removed the sewage from cesspools at night and were able to sell it on to farmers, who used it, in

Opposite:
An illustration from *Punch* showing Michael Faraday confronting the filthy symbol of London's river, Father Thames, during his 1855 trip on the Thames to test the condition of its water.

7

The perils of a defective system of drainage as illustrated in the *Builder* in 1862, showing an overflowing cesspool directly beneath the floor of a wash house.

Map of London showing the courses of its ancient rivers and other watercourses, the majority of which were built over and turned into sewers in the nineteenth century.

modified form, as an agricultural fertiliser.

Ironically, another factor that contributed to the deterioration in London's sanitary state in the early nineteenth century was the improvement in the supply of the city's drinking water. In this period, London's private water companies replaced many of the city's medieval wooden water pipes with equivalents in cast iron, thus enabling them to deliver a more efficient supply at higher pressure. With the simultaneous invention and subsequent popularity of the flushing water closet, much greater volumes of water and sewage were being discharged

into London's rivers and its existing sewers than had previously been the case.

Perhaps the best description available of London's old sewers is that penned by the journalist Henry Mayhew (1812–87) who, in his exhaustive series of books *London Labour and the London Poor* (first collected in the later 1840s and published in four volumes in 1861), investigated the city's underground spaces and interviewed those who worked there, including the nightmen, rat-catchers, scavengers known as

LONDON NIGHTMEN.

[*From a Daguerreotype by BEARD.*]

A pair of London nightmen, as depicted in Henry Mayhew's *London Labour and the London Poor* (1861).

The 'picturesque' rookery of St Giles's as depicted in the *Illustrated London News* in 1849, the year of the second outbreak of cholera in London.

9

An illustration of a rat-catcher in London's old sewers from *London Labour and the London Poor*. Rat-catchers would earn their living by selling their wares to rat-breeders or to disreputable publicans for rat fights.

'toshers', and 'flushers', that is, the workmen who maintained and cleaned the sewers. Mayhew's account contained an encyclopaedic description of the city's sewers – their sizes, lengths and condition – as well as a long and unflinching account of the exact constituents of the London sewage itself. Like many since, Mayhew was both appalled and fascinated by the London sewers, describing the old medieval sewers as 'wretched' and in need of wholesale rebuilding, but also celebrating those who worked in the old sewers in romantic terms, seeing the scavengers and nightmen as nostalgic remnants of a pre-industrial past.

Along with the city's sewers, the slums of London were targeted by sanitary reformers, particularly notorious areas of the city such as the Kensington Potteries and parts of the parish of St Giles's in Westminster. These reformers, who included such luminaries as Charles Dickens (1812–70) and Edwin Chadwick, began to investigate the harmful effects of poor sanitation and to use the evidence they collected – often in the form of lurid descriptions of dilapidated slum housing, streets

Flushing the old London sewers, from *London Labour and the London Poor*.

choked with excrement and festering piles of rubbish – to push the case for immediate reform. Chadwick, in particular, was one of the most obsessive – and belligerent – of all the Victorian sanitary reformers. He used his position as secretary to London's Poor Law Commission (a body established in 1834 to reform London's administrative boundaries and administer welfare to the poor) to focus the attention of the Commission on what Chadwick regarded as the critical relationship between poor sanitation and disease, arguing that those who required welfare were almost always victims of the dirty environments in which they lived. Even popular novels like Dickens's *Bleak House* (1852–53) used realistic descriptions of London's slums to raise awareness of the dangers of leaving poor sanitation unchecked, a move that made these topics common points of discussion in polite society. In *Bleak House* the slum named 'Tom-All-Alone's' was based on the infamous parish of St Giles's in Westminster, with its

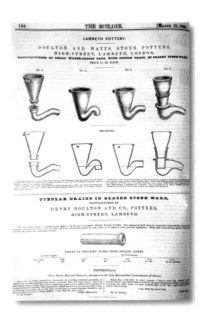

so-called 'rookery' – an area of super-dense housing, with narrow streets, overcrowded tenements and a water supply contaminated with human and animal waste. With this fictional slum, Dickens intended to shock his middle-class readers into an awareness of a very real site of disease in the heart of London's West End that threatened the health of the entire city.

Like many other sanitary reformers in their day, both Dickens and Chadwick believed that disease was generally spread through 'miasma', that is, an infected atmosphere created by a combination of factors: poor drainage, dank and foul air, and gases emitted by diseased bodies. The doctrine of miasma was used by Chadwick and many others to explain the causes of, and solution to, the most feared disease of the nineteenth century: cholera. From the late 1820s, cholera spread rapidly from its original stronghold in Asia across continental Europe, reaching London in 1832 and resulting in 6,536 deaths. Further eruptions of cholera in London were even more catastrophic: the 1848–9 outbreak killed 14,167 people in the city, while that of 1853–54 resulted in 10,738 fatalities. Fear of cholera was particularly intense because it affected all classes of society and

Mid-Victorian advertisement for Doulton & Watts' water closet pans and drainage pipes. Along with Joseph Bramah and Thomas Crapper, Doulton & Watts were perhaps the most famous Victorian manufacturers of sanitary wares.

A pensive Edwin Chadwick, as portrayed in the *Illustrated London News* in 1848, the year he took a leading role in London's new Metropolitan Commission of Sewers.

had a high mortality rate and worryingly rapid development, often killing its victims within a day of onset after acute diarrhea and dehydration. The lack of understanding about the nature of cholera led many people to see the disease as a visitation from a vengeful god, just as the plague had seemed to Europeans in 1348. Despite the pioneering work of Dr John Snow (1813–58), who realised in 1849 that cholera was a water-borne disease, the miasma theory held sway along with an increasing anxiety felt by many about the noxious smells that emanated from the Thames, especially at low tide when raw sewage festered on its mud banks. As a solution, Chadwick and his successor Dr John Simon (1816–1904) pushed for the building of more sewers and the regular flushing of existing ones – well-meant but ultimately misguided policies that only resulted in more and more sewage ending up in the Thames.

The continuing deterioration of London's sanitary infrastructure reached a crisis point in the 1850s, when the toxic smell emanating from the Thames became intolerable, even to the point at which Parliament was about to be relocated upriver to the more salubrious environs of Henley in Oxfordshire. The infamous 'Great Stink' in the long hot summer of 1858 was but the worst of a series of environmental crises in 1850s London.

A magnified drop of Thames water as pictured by *Punch* in 1850. The image, with its mixture of mutant creatures and caricatures of sanitary officials, mocked the complacency of the latter with regard to the state of London's main source of drinking water.

Punch's shocking image of what horrors might emerge from the Thames during the Great Stink in 1858.

Death himself rows along the Thames in search of victims during the height of the Great Stink in 1858.

Throughout this period the satirical journal *Punch* published shocking images of the state of London's once wholesome river, beginning, in 1850, with an image of a magnified 'Drop of London Water', filled with monstrous-looking microbes that satirised those officials who were complacent about the quality of the Thames water. In 1855, *Punch* pictured a journey on the Thames made by the eminent scientist Michael Faraday (1791–1867) to test the quality of the air above its waters, where Faraday exchanges business cards with the monstrously disfigured figure of 'Father Thames'; while, in 1858, during the Great Stink itself, *Punch* imagined the emergence from the Thames of horrific monsters named 'Cholera', 'Scrofula' and 'Diphtheria', as well as Death himself rowing on the Thames ready to strike dead anyone who came near the river. Such images were meant to mock complacency about the possible dangers of the stink and, set against the fear of a new outbreak of cholera, were also a goad to those in power to take urgent action to remedy the problem. In the event, the Great Stink was the final low point in a long history of the environmental degradation of the Thames and, in the summer of 1858, when the smell from the river became intolerable to Members of Parliament, the latter authorised the rebuilding of London's sewers and the Treasury provided 3 million pounds to fund it. The arrival at that decision was the result of many years of patient negotiation and persistence by two very different men: Edwin Chadwick and Joseph Bazalgette.

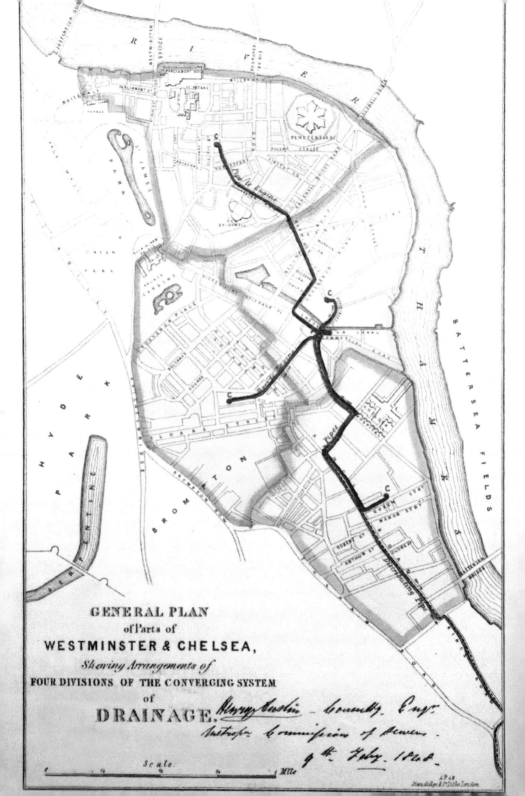

GENERAL PLAN
of Parts of
WESTMINSTER & CHELSEA,
Shewing Arrangements of
FOUR DIVISIONS OF THE CONVERGING SYSTEM
of
DRAINAGE.

Henry Austin — County Eng.

Metropn Commission of Sewers.

9th Feby 1848.

Scale

Standidge & Co Litho London

PLANNING THE NEW SEWERS

IN THE LATE 1840s, Edwin Chadwick was instrumental in the formation of a new governing body for London's sanitation, the Metropolitan Commission of Sewers, which was convened in 1847 following recommendations made by a Royal Commission that had investigated sanitary problems in London. The Metropolitan Commission of Sewers was the first centralised governing body for the city's sanitation, superseding seven of the eight separate sewer commissions that had managed London's sewers since Henry VIII's Sewers Act in 1531 (the City of London Commission remaining separate). From the outset, the main charge of the new Metropolitan Commission of Sewers was to design and construct a unified network of sewers for London. The report of the 1847 Royal Commission had built up a picture of a metropolitan sanitary system on the verge of collapse, a primary cause of disease, governed by a multitude of conflicting interests, and constructed piecemeal from information gleaned from local, parochial surveys. The creation of the new Commission of Sewers was intended as a first step towards the improvement of London's sanitary infrastructure and the abolition of its defective system.

Despite the fact that Chadwick introduced policies that contributed to the deterioration of the condition of the Thames, he also understood very clearly what was necessary in order to begin to plan a new sewer system for London. During his time at the Metropolitan Commission of Sewers he instigated the very first Ordnance Survey of London, arguing that a new map of the city was needed that was grounded in science and the rigorous discipline of the Ordnance Survey. This organisation had its roots in military surveying and, by the mid-nineteenth century, had risen to a position of cartographic dominance over civil surveyors with the production of series of accurate maps of Britain that employed novel surveying techniques drawn from the world of science.

What the first Ordnance Survey of London produced in 1850 were unprecedented maps of the city stripped of its conventional topographic features. The new maps showed only roads and waterways, all carefully

Opposite:
A plan by the engineer Henry Austin to collect the sewage of Westminster and Chelsea for distribution to farmers outside the city, for use as an agricultural fertiliser.

15

Two soldiers from the Corps of Royal Sappers and Miners carry out the first Ordnance Survey of London in 1848 from their observation platform on top of St Paul's Cathedral.

Index to the Ordnance Survey maps of London, produced in 1850 and showing the arrangements of the individual sheets of the 12-inch maps (larger rectangles) and 5-feet maps (smaller rectangles).

measured and marked with levels and benchmarks, the latter corresponding to marks on the ground that still remain to this day. All 901 separate sheets of the Ordnance Survey map were precisely linked together, with the systematic levels and benchmarks providing, for the first time, a unified foundation for calculating the gradients necessary for building a citywide sewer system. This was very much an idealised vision of London because, at the time the Ordnance Survey was carried out, the city had no centralised government: some three hundred bodies administered London in the late 1840s, comprising innumerable parishes and wards, each with their own respective areas of jurisdiction. Despite being the first centralised governing body for London's sanitation, the Metropolitan Commission of Sewers had no real powers either to cut through this administrative confusion or raise the necessary money to

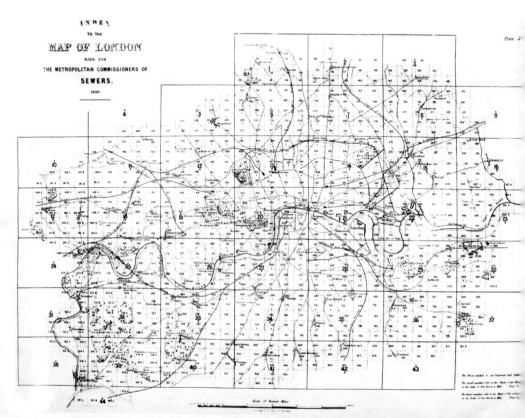

build a citywide sewer network. Only with the creation of the Metropolitan Board of Works in 1855 did the construction of a new sewer system become viable and, even then, it was delayed by much financial and political wrangling.

While the Ordnance Survey was being carried out on the streets of London by around 250 soldiers from the Corps of Royal Sappers and Miners, another survey was going on beneath the

The *Illustrated London News* of 1848 showed a small crowd of onlookers watching a soldier from the Ordnance Survey cutting a benchmark into the ground to indicate where the altitude had been measured.

ground. Chadwick was once again the instigator of this subterranean survey and one of his assistant surveyors was no less than Joseph Bazalgette himself, employed by the Metropolitan Commission of Sewers in 1849 after he had recovered from a nervous breakdown brought on by overwork on the railways in the 1840s. An ignominious beginning to a noble career perhaps, but Bazalgette's first work at the Commission of Sewers gave him the chance to learn, in great detail and at first hand, about the state of London's existing sewers with all their undoubted defects. With his fellow surveyors Joseph Smith and Henry Austin, Bazalgette painstakingly mapped the city's

underground labyrinth in the pages of pocket notebooks that survive in their thousands at the London Metropolitan Archives. Nothing on this scale had ever been attempted before (or since) and it is a testament to the tenacity of Chadwick that it went on for so long: even in 1855, seven years after it had begun, the survey was still continuing. What the surveyors found confirmed their worst suspicions: a system of sewers on the verge of collapse, with rotten brickwork, dangerous blockages and defective design. Unsurprisingly, much of the work was dangerous. In one accident — an explosion of gas in the

Page from one of over a thousand notebooks filled in by the subterranean surveyors of London's sewers. This drawing shows part of a sewer under Lillington Street in Pimlico that was surveyed on 2 September 1848.

Benchmark from
the first Ordnance
Survey of London,
today in Smith
Square, Westminster.

Kennington Road sewer – men working
in the fetid space had the skin peeled off
their faces and their hair singed, while
in another, a surveyor had to be dragged
out of a sewer on his back through
2 feet of black fetid deposit after being
rendered unconscious by the terrible
air inside the sewer. Much of the
information gleaned by the subterranean
surveyors was eventually mapped onto the larger sheets of the Ordnance
Survey, with some experimental sheets also produced at an extraordinary
scale of 10 feet to 1 mile, or double the scale of the Ordnance Survey.

Extract from one
of the experimental
sheets of the
10-feet to 1-mile
survey, showing
the area around
Smith Square in
Pimlico, which was
intended to include
the information
collected by the
subterranean
surveyors.

After Bazalgette's heroic initiation into the topography of London's sewers
he began to rise up through the ranks of the Metropolitan Commission of
Sewers throughout the organisation's many incarnations in the early 1850s.
Periodically formed and reformed, the Metropolitan Commission of Sewers
suffered from intense – and often vitriolic – disputes between its members on
how best to design a new sewer system for London, which resulted in the
premature resignation and subsequent death of its first chief engineer, Frank
Forster. Some of the Commission's members, like Chadwick, believed
passionately in creating a perfect system of sewers right across London, from
larger drains right down to the connections to individual households.

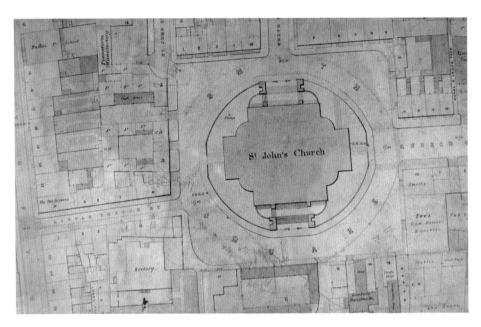

Chadwick's primary motivation was not only to remove all bad smells from the city but also to eventually recycle human waste for use as an agricultural fertiliser; this would be achieved by separating rainwater from waste and then diluting the latter and using it to irrigate farmers' fields.

Chadwick's obsessive interest in developing ways of using human waste as agricultural manure was part of a growing interest in Britain and other countries in the 1840s that focused on the possibility of turning waste into profit. Chadwick believed that the population explosion caused by the Industrial Revolution was only sustainable if human waste could be recycled so that agricultural production could keep pace. Yet, as was proven in several experimental schemes overseen by Chadwick, recycling human waste successfully was immensely difficult to achieve, mainly on account of the enormous expense required both to transport and effectively dilute that waste. Many high-profile recycling schemes in London proved, in the end, to be both impractical and unprofitable. Civil engineers like Forster and Bazalgette took a very different view from Chadwick, favouring a pragmatic approach that diverted all sewage (both rainwater and human waste) out of London through large intercepting sewers and then flushed it into the Thames far enough downstream to avoid it being carried back into the city by the tide. For Bazalgette, the intercepting scheme represented a practical solution to London's sanitary crisis; for others, like Chadwick, it was a profligate waste of a valuable resource.

Bazalgette's predecessor at the Metropolitan Commission of Sewers, the engineer Frank Forster, was the first to develop the concept of an intercepting system of sewers for London, as shown in his 1851 plan for drainage north of the Thames.

In the event, Bazalgette's solution would win out over Chadwick's but not before a host of other schemes had been considered by the Metropolitan Commission of Sewers, including a proposal to carry London's sewerage in an gigantic elevated pipe dozens of miles across the Kent marshes in order to reclaim land in the Thames estuary, or Henry Austin's plan to collect the city's sewage to be distributed to farmers in the countryside around London. When Bazalgette took over as chief engineer to the Commission of Sewers in 1852, he inherited plans from his predecessor Frank Forster and set about developing these into what would become his intercepting system of sewers. With protracted internal wrangling within the various incarnations of the Commission of Sewers, Bazalgette's plans were slow in developing; it wasn't until the Commission was replaced with the Metropolitan Board of Works in 1855 that his scheme was fully approved, the definitive versions of which were drafted in 1856.

Bazalgette explained his scheme by means of two maps attached to his reports. He divided the southern area of London into two catchment areas: the high-level area (coloured pink), to be drained by gravitation alone; and the low-level area (blue), to be drained by a combination of gravity and

Bazalgette's definitive plan for an intercepting system of sewers south of the Thames was presented to the Metropolitan Board of Works in 1855. As is clear in this map, Bazalgette made careful use of the Ordnance Survey of London, carried out in the late 1840s.

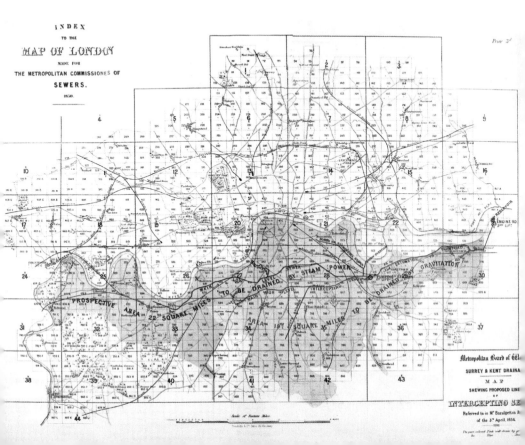

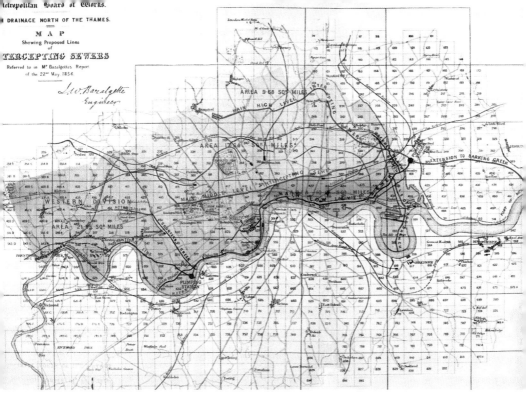

pumping power. Within these areas, two main intercepting sewers – the low- and high-level sewers – would be constructed; the pumping station for the low-level area being located at Deptford Creek – the point where both sewers joined. Bazalgette also included a second pumping station at the outfall at Crossness in the low-level area, owing to the marshy land through which the outfall sewer would pass. Bazalgette's proposal for the drainage of south London was essentially a much-enlarged version of Forster's earlier scheme, but anticipating a much larger future population growth than Forster had done.

With respect to the northern drainage, Bazalgette specified four drainage divisions: the high-level area (yellow), the middle-level area (pink), the low-level area (blue) and a new western division (brown and green). Like those in his plan for south London, Bazalgette's northern intercepting sewers followed similar courses to those originally proposed by Forster, but with extensions in the western suburban districts of the city, which had grown rapidly in the first half of the 1850s. Three main intercepting sewers – the low-, middle- and high-level sewers – would drain the northern areas, the pumping stations being located in Pimlico and West Ham, the latter near to the outfall at Barking Creek. In other sectional drawings, Bazalgette specified the gradients, sizes and shapes of his proposed sewers, favouring an

Bazalgette's plan for intercepting sewers north of the Thames, as presented to the Metropolitan Board of Works in 1856.

egg-shaped cross-section for the smaller sewers, one that Bazalgette thought would make his sewers self-cleansing. The largest of the sewers – those that fed into the outfalls – were circular in cross-section and enormous (in the case of the northern middle-level sewer, up to 12 feet in diameter, or, in today's terms, large enough to accommodate a London bus). Despite protracted wrangling with government officials about where best to locate the outfalls, Bazalgette's scheme was eventually approved by the Board of Works and the government and, partly as a result of the overpowering stench of the Great Stink in 1858, was finally given financial backing from the government, amounting to 3 million pounds.

What was extraordinary about Bazalgette's scheme was both its simplicity and its level of foresight. In effect, he designed a system that would accommodate London's anticipated population growth predicted, by Bazalgette, to rise by 50 per cent from 3 million to a maximum of 4.5 million. This, of course, would prove to be a conservative estimate, with London's population peaking at 8.6 million in 1941, and in the late nineteenth and early twentieth centuries, Bazalgette's intercepting sewers had to be supplemented with new ones in order to increase their capacity. These developments were accompanied by protracted debates about how best to dispose of London's sewage.

From the 1880s onwards, the development of increasingly sophisticated methods of treating sewage at the outfalls at Crossness and Barking meant that the question of sewage disposal lost its previous sense of urgency. However, the dumping of sludge (the part of sewage that remains after treatment) in the North Sea continued until 1998, when European directives came into force that prohibited such methods of disposal. Incineration of the London sludge is now carried out at Barking and Crossness, with experiments continuing on how best to dispose of the highly toxic residual ash. Even today, Bazalgette's main drainage system, although greatly enlarged and badly in need of renovation, remains a primary focus for debates on the future of London's sanitary infrastructure. The proposed Thames Tideway Tunnel,

Bazalgette's sectional drawing of the northern low-level intercepting sewer, showing its entire course underground by means of a 'squashed' section, that is, with widely differing horizontal and vertical scales. Above the section is shown the progressive growth in the size of the sewer, as well as its transformation from an egg to a circular shape in cross-section.

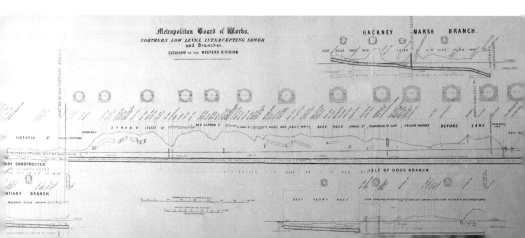

The graceful outlines of Crossness's sludge incinerator, built by an AMEC-Lurgi consortium and opened by the Duke of Edinburgh in November 1998.

a giant 7.2-metre diameter 'supersewer' to run 24 miles deep beneath the bed of the River Thames from Hammersmith to Beckton, is being touted by Thames Water plc as the 'second phase' of Bazalgette's intercepting system. Designed to prevent the regular overflow of Bazalgette's system into the Thames (which continues to pollute the river and cost Thames Water millions of pounds in annual fines), it remains to be seen whether the supersewer will be approved and then completed by its estimated date of 2021 and within its projected budget of £3.6 billion.

Map of the proposed routes of London's new 'supersewer', the Thames Tideway Tunnel, to be built to enlarge the capacity of the city's sewers so that raw sewage no longer enters the Thames. The scheme is in fact very similar to one rejected by Bazalgette in 1849.

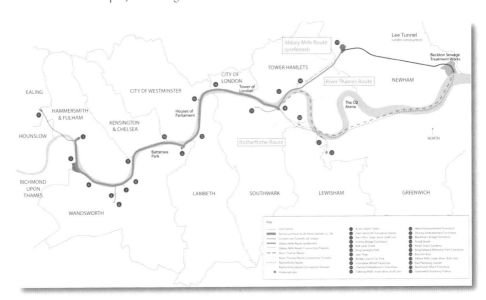

CONSTRUCTING
THE NEW SEWERS

BAZALGETTE's plans were finally approved in the summer of 1858, and construction began in earnest. In all, twenty-seven contracts were drawn up for the first phase of the building of the main drainage system, all of which were tendered from 1859 to 1865, with many running simultaneously. With such an enormous project, comprising over 82 miles of intercepting sewers, Bazalgette tended to favour using reliable contractors like George Furness and William Webster rather than giving the contract to the lowest tender – a practice that would later lead to accusations of favouritism on the part of Bazalgette and which contributed to the downfall of the Metropolitan Board of Works in 1888. An idea of the extent of the main drainage contracts can be gained from the original drawings held at the London Metropolitan Archives. For example, the contract for the northern outfall sewer, drawn up from 1859 to 1860 and awarded to Furness, was made up of fifty-one large-scale drawings and an eighty-five-page printed specification outlining the estimated quantities and costs of materials. The drawings were produced by an army of draughtsmen (over four hundred for the twenty-seven contracts) working unrestricted hours to tight deadlines. As with all the original contracts, each of the northern outfall sewer drawings was meticulously hand-coloured; they included plan and sectional views of the course of the outfall sewer and details of obstacles to be negotiated, such as roads and rivers. Unlike the majority of the main drainage system, the 4-mile-long northern outfall sewer was built entirely above ground, the treble-line of sewer tunnels enclosed in a 30-foot-high embankment because of the low-lying and marshy nature of the ground over which they passed. Even today, that embankment, enlarged in the 1920s and now known as the Greenway, still remains a formidable obstacle to any future redevelopment of the area of Hackney that it straddles.

Victorian contractors like Furness were adept at managing an enormous workforce, most having cut their teeth in railway contracting in the 1840s, which often involved managing many projects simultaneously. Hundreds of

Opposite:
The navvies as heroes, as seen in this 1859 engraving from the *Illustrated London News* showing the construction of the junction of the northern high- and middle-level sewers at Wick Lane in Hackney.

individuals would have built the northern outfall sewer alone and these included general excavators (known as 'navvies'), bricklayers, surveyors and concrete-mixers. Meanwhile, Bazalgette would have provided his own workforce of supervisors: on-site engineers and clerks of works, the latter drawing up regular statistical reports for Bazalgette on the progress and costs of construction.

The top of the embankment that carries the five pipes of the northern outfall sewer within it. Today, the embankment is called the Greenway, a cycle and pedestrian route that was used as an access route to the 2012 Olympic site in Stratford.

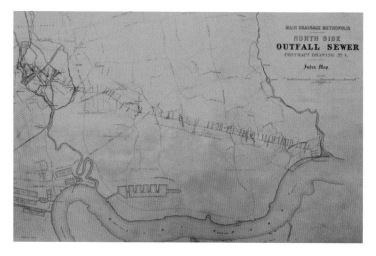

Contract drawing from 1860 showing the proposed course of the northern outfall sewer: 4 miles across the marshy ground east of West Ham to the outfall at Barking Creek.

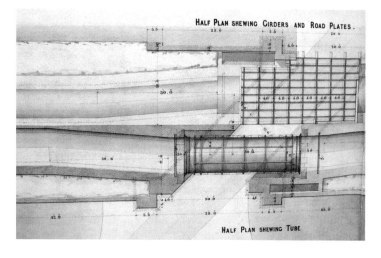

Extract from an 1860 contract drawing for the northern outfall sewer showing a proposed viaduct to carry the sewer across a road.

Despite many problems encountered during the building of the northern outfall sewer — the 1859–60 labourers' strike over working hours, continuous problems with heavy summer rains, hard winter frosts and difficulties obtaining enough brick and stone — the project was completed by Furness on time in just two years. Bazalgette was not as fortunate with some of his other chosen contractors: William Rowe, who was awarded the contract for the northern middle-level sewer in February 1860, went bankrupt soon after starting work and the contract was eventually completed by the notable Victorian railway contractor Thomas Brassey (1805–70). In the light of such failures, it was no wonder that Bazalgette ended up favouring reliable contractors, even if this tarnished his professional reputation.

The construction of the main drainage system in London in the 1860s occurred alongside that of many other significant engineering projects in the capital, including the Metropolitan Railway, the world's first underground railway, the first section of which was completed in 1863; the London, Chatham and Dover Railway (from 1862), which drove the first

Photograph from 1862 showing the construction of the northern outfall sewer (shown in the background). On the left in the foreground is George Furness's cement-mixing machine; on the right, Bazalgette, his on-site clerk of works and Furness consult a contract drawing.

Sectional view of the Victoria Embankment near Charing Cross railway station, showing the proposed infrastructure, including: (1) a subway for water and gas pipes and telegraph cables; (2) the northern low-level sewer; (3) the Metropolitan Railway; and (4) the pneumatic railway powered by compressed air.

line of railway over the Thames into the heart of the City of London; new street improvements such as Holborn Viaduct (1866–9); and the Thames embankments, begun in 1862. The embankments were both a significant engineering project in their own right and also critical in relation to the completion of the main drainage system; housed within the walls of the

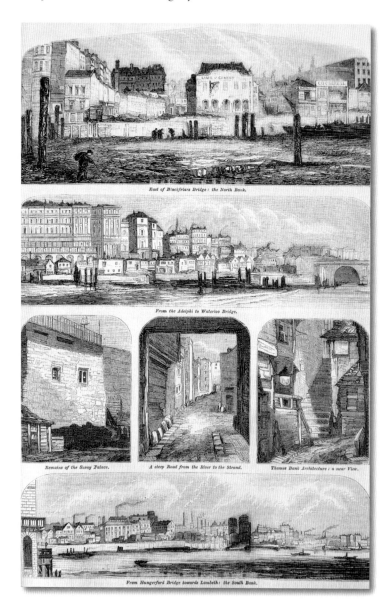

East of Blackfriars Bridge : the North Bank.

From the Adelphi to Waterloo Bridge.

Remains of the Savoy Palace.　　A steep Road from the River to the Strand.　　Thames Bank Architecture : a near View.

From Hungerford Bridge towards Lambeth: the South Bank.

Views of the Thames just before embanking began, showing scavengers on its mud banks and the variety of industrial and commercial premises that once lined its shores.

Victoria Embankment (1864–70) was the northern low-level intercepting sewer. Just like the main drainage system, the embankments were conceived against a background of concern about the sanitary state of the Thames.

Before embanking began in 1862, the banks of the Thames were lined with various industrial and commercial premises: wharfingers whose livelihoods depended on direct access to the river but who generally contributed to the latter's poor sanitary condition, which drew so much criticism in the 1850s. By reclaiming 52 acres of land either side of the Thames, the embankments not only swept away these former industries (consigning most of them further downriver); they also enabled an entirely new river frontage and roadway to be constructed, one that forever altered the image of the city as a whole. Less obvious, but no less important, was the honeycomb of tunnels inside the embankments, including, in the Victoria Embankment, a shallow tunnel for a new section of the underground railway (now part of the District and Circle lines); a tunnel for water, gas and, later, electricity cables; and a separate tunnel for the low-level sewer. Such was the perceived national significance of the embankment project that, when the completed Victoria Embankment was opened on 13 July 1870, it was done so with great formal ceremony, with the Prince of Wales leading a procession of other royals and dignitaries, but without Queen Victoria who was apparently too ill to attend. Much was made of the Victoria Embankment as the London

The Victoria Embankment today, looking west towards the Houses of Parliament.

One of the many construction sites in London in the early 1860s: the new line of the London, Chatham and Dover Railway that was the first to cross the Thames at Blackfriars.

Elevated view of the Victoria Embankment on the occasion of its ceremonial opening in 1870.

equivalent of Paris's new boulevards; many newspaper accounts celebrated its making London more magnificent (and cleaner) than its French rival. In recognition of this, Bazalgette was eventually knighted by the Queen for his work on both the main drainage system and the embankments. His memorial, unveiled in 1901, was placed, appropriately enough, on the walls of the Victoria Embankment near to what is now the Embankment Underground Station.

Monument commemorating the life of Sir Joseph Bazalgette, situated on the Victoria Embankment, near Embankment Underground Station.

Throughout the construction of both the main drainage system and the Thames embankments, London's press showed a great deal of interest, especially as Londoners themselves would eventually have to pay for these projects by means of a 3p rate imposed for forty years on all properties within the metropolitan area. Today, we take for granted, and usually resent, the disruption caused by urban improvements such as the excavation of roads and the repair or upgrading of infrastructure. However, in the 1860s, London's populace had never before witnessed construction on this industrial scale in their city; the sheer vastness of the spectacle overawed and, at times, overwhelmed the public. We have no better visual record of the construction of the main drainage system than that presented in the pages of London's leading pictorial periodical, the *Illustrated London News*. Founded in 1842 as the world's first illustrated newspaper, the *Illustrated London News* placed great emphasis on depicting everyday news stories, and particularly ones that related to the improvement of the capital's condition. In this context, for the *Illustrated London News*, the rebuilding or repair of London's sewers was not simply a mundane event; rather, it was something to be celebrated as progressive and brought to the attention of middle-class readers who might not have been aware of it. Thus, in its 1845 engraving of the deepening of the Fleet sewer – one of London's old sewers that enclosed the ancient River Fleet – the *Illustrated London News* dramatised the event with an image that showed, with a palpable sense of wonder, the great depth of the trench dug beneath Fleet Street to deepen and enlarge the capacity of the sewer.

A dramatic engraving in the *Illustrated London News* showing the deepening of the Fleet sewer in 1845.

Unsurprisingly, the construction of Bazalgette's main drainage system from 1859 onwards provided the *Illustrated London News* with a great deal of illustrative subject matter that it could enthusiastically embrace. From the start, the newspaper drew attention to the sheer scale of the project, referring to the use of 318 million bricks, 880,000 cubic yards of concrete and the excavation of 3.5 million cubic yards of earth. The *Illustrated London News* also pictured every part of the building process: the first trench dug for the northern high-level sewer near Victoria Park in February 1859; the commencement of the vast northern outfall sewer at Wick Lane in Hackney later in the same year; an array of different sites visited by the newspaper and other members of London's press in 1861; and the final stages of the construction of the outfall at Barking in 1864. Together, these images showed a project advancing in its scale and scope, from an early emphasis on the forms of the sewers themselves to later images depicting vast construction sites often from an elevated viewpoint, which emphasised

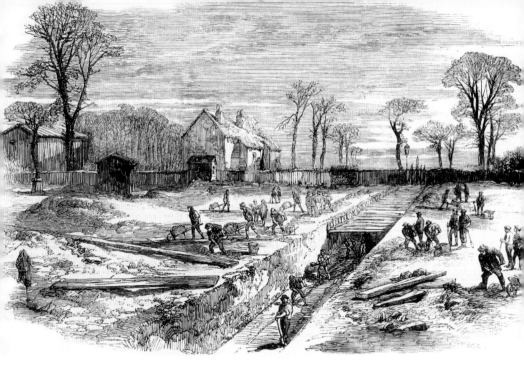

Starting the construction of London's main drainage system near Victoria Park, Hackney, as pictured by the *Illustrated London News* in 1859.

their scale. After the press were guided by Bazalgette around the construction sites in 1861, many newspapers opined that the main drainage system was equivalent to, and even surpassed, the seven wonders of the ancient world. For the press, London's sewers were a modern technological wonder, and superior to their ancient equivalents because they were guided by liberal, rather than despotic, politics.

The images of the construction of the main drainage system in the *Illustrated London News* were also designed to inform the newspaper's readers (generally upper-middle-class Londoners) of the beneficial nature of the project and to reassure them that the chaos brought to the streets of London by the excavation work was guided by a noble aim, efficient engineering principles and competent supervisors. In addition, through its many engravings of the construction process, the *Illustrated London News* helped respectable middle-class Londoners come to terms with the presence of thousands of navvies in London during the 1860s – lower-class workers who were generally perceived to be a source of foul language, drunkenness and other social vices. The engravings presented the navvies in a positive, even heroic, light, celebrating them as essential to the realisation of the project.

Yet, inevitably, during the construction process there were accidents: events that brought to light the more destructive power of such a vast engineering project. One of the first serious accidents occurred on 28 May

1862, when an explosion at Shoreditch – caused by a ruptured gas main next to the excavation for the northern high-level sewer – resulted in localised destruction and the death of an unfortunate passer-by. The gas that escaped, according to the *Illustrated Weekly News*, 'rushed out with a noise resembling a perfect hurricane' before which men were floored 'instantly'. On 18 June 1862, a second and much more destructive accident attracted

Constructing the northern outfall sewer near the outfall at Barking, which was witnessed by representatives of London's press in October 1861.

The vast scale of the construction works at the Barking outfall, as depicted by the *Illustrated London News* in 1864.

TRINITY   HIGH   WATER

GROUND

the attention of the majority of London's illustrated newspapers: the bursting of the walls of the Fleet sewer in Victoria Street, Clerkenwell. The accident occurred as a result of the close conjunction of two major building projects: the Metropolitan Underground Railway and the main drainage system, which encompassed the rebuilding of the Fleet sewer. Both projects involved the excavation of deep trenches adjacent to one another and after heavy rain on 18 June, the Fleet sewer flooded, which in turn inundated the railway trench and eventually the entire surrounding district. The resulting spectacle was given a front-page engraving by the *Illustrated London News*, showing the 'black hole' of the Fleet sewer, the damaged walls of the Underground Railway, and exposed gas mains, water pipes and even gravestones, as well as a buried lamp post. A final, more serious, accident happened in Deptford in south London on 29 April 1863, when a section of sewer tunnel collapsed and buried six navvies. After fifteen hours of digging – much of it by hand – only three survivors were hauled out from the mass of earth and timber. Despite these accidents and other setbacks, the building work progressed remarkably fast – a mark of Bazalgette's tenacity and even ruthlessness.

Front-page news for the *Illustrated Weekly News*: the results of a gas explosion during the construction of the main drainage system in Shoreditch in May 1862.

By the spring of 1865, all of the drainage works in south London were completed on time, with those north of the river continuing until 1870, when the Victoria Embankment was opened. However, the £3 million provided by the Treasury had to be supplemented by an additional loan of £1.2 million to cover additional costs during construction. As the project neared completion, attention shifted from construction to plans for the opening ceremonies to be held at two of the main drainage pumping stations at Crossness and Abbey Mills.

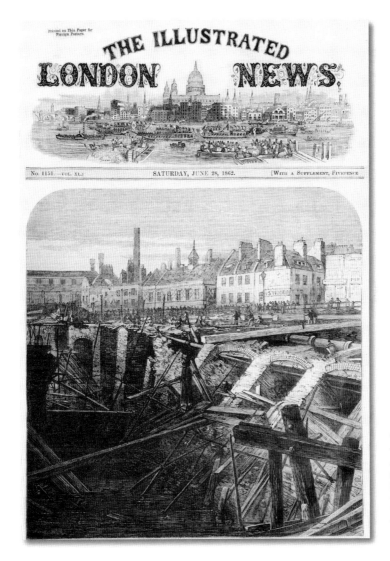

The front page of the *Illustrated London News* depicts the scene of destruction in Clerkenwell when the Fleet sewer burst through its walls after heavy rain in June 1862.

CATHEDRALS OF SEWAGE

THE CROSSNESS (1862–5) and Abbey Mills (1865–8) pumping stations – the largest of the four connected with the main drainage system – were, and still are, vital components of that system. They raise sewage intercepted from low-lying areas of London to allow it to drain by gravitation into outfalls located outside the city limits at Barking, on the north side of the river, and Crossness on the south. The pumping stations that were built as part of Bazalgette's new drainage system were the first of their kind, London's main drainage system being the first urban sewerage scheme to use extensive pumping power. However, in terms of their function, they were identical to other kinds of pumping stations that had been developed out of the earlier needs of the Industrial Revolution, principally to pump water out of deep-level tin mines in Cornwall. The steam engines employed for pumping were enormous pieces of machinery and ones that required tailor-made structures (known as engine-houses) to contain them. After the development of high-pressure Cornish engines around 1812, pumping engines began to be used in the supply of water. The immense size and weight of Cornish engines – some with up to 100-inch cylinders – required large buildings in which to house them, notable examples being the engine-houses at Kew Bridge (1838), Hampton (1855) and Stoke Newington (1856). When conceiving of the pumping stations for his main drainage system, Bazalgette would have been guided by these earlier precedents built for the mining and water industries.

Bazalgette's pumping stations may have been similar in their functionality to those developed for water supply; yet they were very different in terms of their architectural style. Compared with the restrained classicism seen in the Kew Bridge water-pumping station and in the smaller main drainage pumping stations at Deptford (1859–62) and Pimlico (1870–4), the architectural extravagance of Crossness and Abbey Mills was significant. Located on the south side of the Thames, Crossness was a stylistically eclectic building, clearly designed to impress: the engine-house included

Opposite:
The decorative cast-iron octagon in the centre of the interior of the engine-house of the Crossness pumping station.

An early example of a water-pumping-station design can be seen in the engine-house at Kew Bridge (1838).

A water-pumping station in the guise of a medieval castle: Green Lanes, Stoke Newington, was built from 1854 to 1856 and designed by William Chadwell Mylne.

a cathedral-like main entrance, a striking campanile-like chimney (now demolished) and elaborate interior decorative ironwork, the centrepiece of which was the central octagonal structure in a mixture of wrought and cast iron. The design features seen at Crossness were continued and developed at Abbey Mills, where the decorative octagon was transformed into the

A photograph of the Deptford pumping station (1859–62) in the 1950s, showing its twin engine-houses joined by a single-storey boiler house and prominent chimney in the background (now demolished).

building's most striking architectural feature and the internal ironwork was both more unified and more lavishly ornate than that at Crossness. The original twin ventilation chimneys (demolished in 1940), richly ornamented and standing 212 feet high, gave this building a prominence that has consistently attracted public attention; today it still provides a focus

The Western pumping station (1870–4) in Pimlico, as it is today.

The Crossness pumping station pictured in the *Builder* in 1865, showing its two-storey engine house with Mansard roof and ornate chimney resembling an Italian campanile.

for introducing the public to Bazalgette's system.

The design of both Crossness and Abbey Mills was the result of a partnership between Bazalgette and an architect, Charles Henry Driver (1832–1900), their input being focused on the functional and decorative aspects respectively. Driver probably first met Bazalgette when they worked together as assistant surveyors for the Metropolitan Commission of Sewers from 1849 to 1850; at this time many would-be architects began their careers in this way, as surveyors or draughtsmen working alongside aspiring engineers. Driver then left the Commission of Sewers and went to work as an architect for the railways, designing numerous stations in the 1850s and 1860s in London, the Midlands, and the south of England before moving on to more ambitious projects such as the Horton Infirmary in Banbury (1869–72) and the Crossness and Abbey Mills pumping stations. Driver was well known for his expertise in the architectural use of ornamental cast iron, delivering many papers on the subject and developing long-standing

Exterior of Abbey Mills pumping station, as pictured by the *Illustrated London News* in 1868.

partnerships with leading ornamental iron founders such as Walter Macfarlane in Glasgow. Diverging from the views of leading critics such as John Ruskin (1819–1900), Driver argued passionately for the use of new materials like cast iron, in both engineering projects and as architectural decoration. Driver's versatility as an architect did not go unrecognised in his lifetime, as indicated in his obituaries in the building press in 1900. Yet, because he often worked with engineers, his significance as a Victorian architect remains difficult to assess; in Driver's day, engineers were the heroic constructors and, in several cases, became far better known than architects of similar stature.

In relation to the design of the Crossness and Abbey Mills pumping stations, Driver was employed as an architectural consultant by Bazalgette. At Crossness, Driver probably provided the designs for the flamboyant interior ornamental cast ironwork, while at Abbey Mills he contributed to the decoration of the entire building, both inside and out. Even by Victorian standards, Abbey Mills is an extraordinary building, bringing together a wide array of stylistic features: Italian and English Gothic windows, Islamic chimneys, Byzantine ironwork, a French roof, a Greek Orthodox cross-shaped floor plan and Renaissance doorways. The fantastical appearance of the building led some contemporaneous commentators to describe it as resembling 'a mosque in a swamp', a 'Chinese temple' or a 'cathedral of sewage'. No doubt, the building was intended to impress, making a visual statement that would rise above its prosaic and rather unseemly function; but it was also designed to give a sense of what a new style of architecture might look like, one that was tailored to brand-new building types such as sewage pumping stations. In addition, the pervasive naturalism of the ornamentation that adorns the pumping station – particularly in the frieze of British flora that runs right around both the

Interior of the Abbey Mills pumping station today, showing the liberal use of ornamental cast iron by the architect Charles Driver.

Contemporary view of the Abbey Mills pumping station from the top of the Greenway.

Charles Driver's
Battersea Park
railway station, built
in 1866 and mixing
Italian (round-
arched windows)
and French
(segmental arched
first-floor windows)
styles in its three
storeys.

The cast-iron
railings at Abbey
Mills include striking
roundels, containing
the coats of arms of
the main sponsors
of the primary
drainage system,
surrounded by
extravagant lilies.

interior and exterior of the building – suggests a more literal connection
between sewage and natural abundance. For, like many Victorians, Driver
was a passionate enthusiast for the recycling of
sewage as an agricultural fertiliser. However,
Bazalgette's main drainage system did not include
any process by which London's waste might be
turned to profit; by the late 1860s, following the
failure of many high-profile schemes intended to
recycle London's sewage, most enthusiasts had given
up any hope of achieving this. Yet, in the abundant
naturalism of Abbey Mills, there was still the hope
that this may yet be achieved, that is, a sustainable
way of living based on the unification of man and
his waste.

Even if most visitors to the Abbey Mills pumping
station did not pick up on Driver's association of
sewage with natural abundance, there was a general
sense in which the lavishness of the decoration of
both Crossness and Abbey Mills was seen as a suitable
reflection of their role as symbolic representatives
of London's main drainage system as a whole,
particularly as most of that system lay invisible

beneath the city. The opening ceremonies held in 1865 at Crossness and 1868 at Abbey Mills confirmed the important symbolic role of the pumping stations; both marked the operational starting of the main drainage system south and north of the Thames respectively. The Metropolitan Board of Works used these ceremonies to promote their prowess as urban planners, whether to Members of Parliament, archbishops, foreign dignitaries, or royalty – the Prince of Wales being asked to open Crossness. Many hundreds of guests attended the ceremony at Crossness, held on 4 April 1865, with special trains laid on from Charing Cross to the remote site on the Essex marshes and a steamboat carrying the royal party from Westminster. The events of the day included tours of Crossness's underground sewage reservoir, an explanatory lecture by Bazalgette, a ceremony in the lavishly decorated engine-house (where the Prince of Wales started the engines), and a lavish banquet in one of the workshops.

Invitation to the opening ceremony at the Crossness pumping station, sent to over six hundred guests in 1865.

The ceremony at Abbey Mills on 31 July 1868 followed a similar, if stripped-down, schedule to that at Crossness. Visits to Abbey Mills also continued after the main ceremony: during the following weeks, representatives from London's vestries (the Victorian equivalent of local councils) visited the pumping station in a succession of organised tours, with many of the city's local newspapers giving their verdicts on the significance of the new drainage system. The majority celebrated the cleansing of the metropolis, even if a few of them grumbled at the cost of the pumping station, reflecting the fact that it was the ratepayers of London who would ultimately pay for the city's new drainage system.

The many press accounts of these ceremonies gave expression to a range of responses to the new main drainage system: reports of its technical details, drawn from Bazalgette's own descriptions; paeans of wonder at its massive

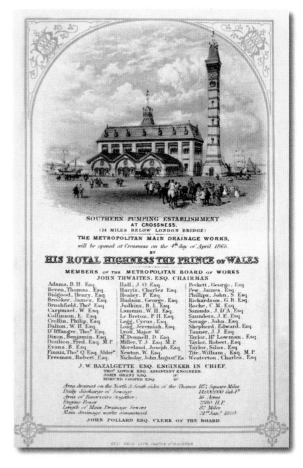

scale and noble function; and a sense of disquiet at the monstrous and unprecedented quantities of sewage it now discharged into the Thames. One aspect that many of the press articles stressed was the relationship between the architectural style of the buildings and their function, namely, the pumping of sewage. The ceremony at Crossness in 1865 in particular provoked strong reaction. The *Standard* thought that an 'enchanter's wand' had touched the whole site at Crossness, particularly the interior of the engine-house which seemed to the newspaper to be a 'perfect shrine of machinery'. The *Daily News* took up this religious imagery, describing the 'beautiful octagon' in the centre of the engine-house as resembling the interior of a Byzantine church, with the steam engines themselves taking on Christian connotations, one of the great cylinders becoming a 'pulpit'. In a more general sense, newspaper responses to Abbey Mills referred to the 'tremendous engines' seen in the engine-house, which generated a sense

Photograph of assembled dignitaries outside the Crossness pumping station in 1865.

of 'deep wonder and admiration' when surrounded by the abundant ornamental ironwork.

During the ceremony at Crossness, male visitors were invited to descend into the crypt-like space of part of its vast subterranean sewage reservoir. Despite the temporary exclusion of the sewage and the dazzling lighting, some visitors felt distinct unease at the thought of being in such close proximity to what the writer for the *Daily Telegraph* termed 'the filthiest mess in Europe', ready to 'leap out like a black panther' after the guests had left. Even if the sewage was invisible, it was present in the visitors' imagination; they felt invaded by an unseen danger and placed 'in the very jaws of peril, in the gorge of the valley of the shadow of death', separated only by bolted iron gates from the 'pent up and bridled in' sewage. As if to exemplify the unique character of this experience, three years later, in the ceremony at Abbey Mills, visitors also wondered at the lavish decoration and vast machinery but did not refer to any monstrous associations. Indeed, when presented with the opportunity of inspecting the sewage pumps below ground, most of the visitors declined; even the *Daily Telegraph*, whose correspondent had, three years earlier, been so rampant in his imaginative prose, gave little attention to these 'noisome chambers far below' the building's sumptuous interior.

Below left: The Prince of Wales turns on the steam engines in the interior of the engine-house at Crossness, as seen on the front page of the *Illustrated London News*, 15 April 1865.

Below: Page layout from the *Illustrated London News*, 15 April 1865, showing Bazalgette giving a lecture on the main drainage system (upper engraving) and the banquet in one of the workshops adjoining the engine-house (lower engraving).

Visitors admiring the decorative ironwork inside the engine-house of the Crossness pumping station in 1865.

The visit to the subterranean sewage reservoir at Crossness in 1865 would be a unique opportunity to see the underground parts of London's new sewer system; thereafter, the sewers remained generally sealed off to curious visitors. This was in direct contrast to the situation in Paris, where the city's new sewers, constructed at the same time as those in London, were opened for public tours during the Exposition of 1867, and have remained so to this day, now as part of the *Musée des Égouts*, situated on the banks of the Seine near the Eiffel Tower. Initially, male and female visitors to the Paris sewers rode in tailor-made boats through the Parisian sewage itself which, despite being absent of human excrement until the end of the

Male visitors being instructed by Bazalgette inside the subterranean sewage reservoir at Crossness.

Visitors being taken on a tour of the Paris sewers in 1870, showing women riding in boats.

nineteenth century, was nevertheless still regarded as being as noisome and filthy as London's was in 1865. Yet, the first tourists of the Paris sewers were reassured by the orderliness of the new spaces and by the bright lighting; male visitors were entranced by the presence of ladies dressed in their finery and also imagined themselves in the role of the chivalrous hero of Victor Hugo's *Les Misérables*, published in 1862, where, in the climactic moments of the epic novel, the hero Jean Valjean makes his escape in the Paris sewers after the 1832 revolution. In direct contrast to Paris, it seemed that, after London's sewers were largely completed in 1868, they were destined to remain sealed off forever.

Contemporary view inside the Paris sewers; part of the city's *Musée des Égouts*.

EXPLORING LONDON'S SEWERS

AFTER THE GRAND OPENING of the main drainage system in 1865 and 1868, London's new sewers continued to be refashioned – principally the smaller street sewers and household drains that all had to be connected up to the intercepting system (a process initially overseen by Bazalgette himself). There was also the continuing question of how best to deal with the sewage, which, in Bazalgette's intercepting system, was carried away from London's main built-up area and simply dumped into the River Thames further downstream. Unsurprisingly, the Thames Conservatory Board raised repeated objections to the vast amounts of sewage that were now polluting the river around the outfalls at Barking and Crossness. These disputes were intensified after a horrific accident in September 1878, when the pleasure steamer *Princess Alice* sank near the outfalls after a collision with a freighter, resulting in over six hundred deaths, which were mostly the result of poisoning by sewage rather than drowning. Later, in the 1880s, there was increasing concern that London might experience another 'Great Stink' on the scale of that witnessed in 1858, and this finally led the Metropolitan Board of Works to begin purifying the sewage at the outfalls instead of flushing it untreated into the Thames.

Away from the outfalls, London's sewers continued to be enlarged as the population of the city grew far beyond that envisaged by Bazalgette; yet their engineering basis remained the same, with new lines of intercepting sewers added to increase the capacity of the original ones. Despite cleaning up the city, the intercepting sewers were never a perfect system; in times of heavy rain, raw sewage was, and still is, flushed into the Thames to avoid overloading the sewers.

As in Bazalgette's day, London's sewers continue to fascinate, despite (or even perhaps because of) their invisibility. In their annual 'Open Sewers Week', Thames Water plc open up the original Abbey Mills pumping station, showing visitors its elaborate decoration and pumping engines

Opposite:
Walking through
the northern
outfall sewer during
Thames Water's
'Open Sewers
Week', May 2007.

(albeit electric versions of the original steam-powered ones). The event also includes an optional underground walk through a small section of the northern outfall sewer at Wick Lane in Hackney. After being kitted out in protective clothing, lead-weighted boots and an oxygen tank in case of poisonous gas, visitors are lowered by a rope into the cavernous space of the northern outfall sewer. Once inside, it is clear just how hostile the interiors of the London sewers are; the visit entails wading through knee-deep sewage along a smooth but slippery brick-lined tunnel in almost complete darkness, guided only by the maintenance team who work in the sewers. For the majority of the public, however, London's sewers can only be experienced remotely, for example, through television adaptations telling their story, such as the BBC film *The Sewer King* (2003), part of its series *The Seven Wonders of the Industrial World*, or through museum exhibits, such as the reconstruction of a London sewer at the Kew Bridge Steam Museum, complete with a peephole view of a London sewer.

The restrictions on access to London's sewers have not deterred those determined to investigate these spaces, albeit illegally. Perhaps one of the first illicit sewer explorers was the Victorian journalist John Hollingshead

One of London's dwindling band of sewer maintenance workers, pictured inside the northern outfall sewer, May 2007.

A view into one
of London's sewers:
the peephole exhibit
at the Kew Bridge
Steam Museum.

(1827–1904) who, in 1862, published *Underground London*, an account of his travels beneath the city, but which mainly focused on London's sewers. Hollingshead had a self-confessed 'appetite for the wonderful in connection with sewers' and his book catalogued many aspects of London's drainage system, including the city's old drains, journeys within the sewers themselves, or Bazalgette's main drainage system, then under construction. As he stated in the introduction to the book:

> There are more ways than one of looking at sewers, especially old London sewers. There is a highly romantic point of view from which they are regarded as accessible, pleasant and convivial hiding-places for criminals flying from justice, but black and dangerous labyrinths for the innocent stranger … [and] there is the scientific or half-scientific way, which is not always wanting in the imaginative element.

In keeping with this view, Hollingshead described London's main drainage system as a 'great accomplished fact' and a successful example of the 'struggle of art against nature'; yet, he also stated that its vast intercepting sewers, which 'dwell in perpetual darkness' and concentrated all of London's sewage in one space, might be seen by some as 'volcanoes of filth; gorged veins of putridity; ready to explode at any moment in a whirlwind of foul gas, and poison all those whom they fail to smother.' In a wonderful moment during one of his sewer journeys, Hollingshead was told by his guide that he was now walking beneath Buckingham Palace, whereupon he promptly sang the national anthem, while up to his knees in what was, presumably, royal excrement.

Hollingshead's intrepid journeys are mirrored today in the practice of draining, an increasingly popular branch of urban exploration. For urban explorers, illicit sites – industrial ruins, abandoned buildings and underground spaces – are places where the normal rules of city life can be challenged. Visiting sewers presents an opportunity to discover a secret world under the city, one that might provide a completely new insight and experience of the city. Usually under the cover of night, sewer explorers descend into these spaces and explore them at will. This usually involves a degree of danger: mobile phone networks cease to operate; the space is usually pitch black, slippery underfoot, and highly disorientating. Because the sewers are designed in a grid-like network, they are easily comprehensible on a map of the city, but not so underground. However, this is unsurprising, for Bazalgette assumed there would be no visitors to these spaces: they were conceived to be self-cleansing and therefore no walkways or other helpful features for people were incorporated into their design.

There has been a variety of imaginative uses of the city's sewers in both television and film. An early example was the giant human-eating rat that hid in the London sewers in 'Gnaws', a memorable episode of the 1976 ITV series *The New Avengers*. More recently, in Neil Gaiman's 1996 BBC television series and popular book *Neverwhere*, London's underground is re-imagined as the fantasy world 'London Below', where mythical characters use the city's subterranean spaces – the sewers, the Tube, even the Crossness sewage reservoir – as passages through a city with a seemingly impossible alternative geography. In a different vein, Clare Clark's 2005 historical novel *The Great Stink* sets most of its Victorian narrative in the London sewers, exploiting their dark associations to mirror the repressed yearnings of her central character, which are played out in the hidden spaces of the sewers before dramatically entering the life of the world above. More straightforwardly horrific is the deformed monster living in a self-made subterranean labyrinth under London in Christopher Smith's 2004 horror film, *Creep*. This monster – the sole survivor of clandestine genetic experiments carried out under London – lives in the city's sewers and in the Underground network, only emerging into the public part of the Tube to carry out brutal murders. Although far from subtle, the horrors in *Creep* seem to chime on some level with the much more tangible unease now associated with the city's substructure, particularly in the immediate aftermath of the terrorist attacks of 7 July 2005. London's sewers might have originally been built to bring the city's waste to order, and to keep these horrors reassuringly out of sight (and smell); yet, for some, these spaces represent something different: a fascinating, alternative way of seeing the city.

A storm drain beneath Brockwell Park in Brixton.

FURTHER READING

Ackroyd, Peter. *London Under*. Vintage, 2012.

Clarke, Clare, *The Great Stink*. Penguin, 2006.

Clayton, Anthony. *Subterranean City: Beneath the Streets of London*. Historical Publications Ltd, 2010.

Dobraszczyk, Paul. *Into the Belly of the Beast: Exploring London's Victorian Sewers*. Spire, 2009.

Emmerson, Andrew. *Discovering Subterranean London*. Shire, 2009.

Finer, Samuel E. *The Life and Times of Sir Edwin Chadwick*. Barnes & Noble, 1970.

Gaiman, Neil. *Neverwhere*. BBC Books, 1996.

Halliday, Stephen. *The Great Stink of London: Sir Joseph Bazalgette and the Cleansing of the Victorian Metropolis*. Sutton, 1999.

Hamlin, Christopher. *Public Health and Social Justice in the Age of Chadwick: Britain 1800–1854*. Cambridge University Press, 1998.

Harrison, Michael. *London Beneath the Pavement*. Peter Davies, 1961.

Hollingshead, John. *Underground London*. Groombridge & Sons, 1862.

Hyde, Ralph. *Printed Maps of Victorian London: 1851–1900*. William Dawson & Sons, 1975.

Jephson, Henry Lorenzo. *The Sanitary Evolution of London*. Fisher Unwin, 1907.

Jones, Edgar. *Industrial Architecture in Britain, 1750–1939*. B. T. Batsford, 1985.

Lesser, Wendy. *The Life Below Ground: A Study of the Subterranean in Literature*. Faber, 1987.

Luckin, Bill. *Pollution and Control: A Social History of the Thames in the Nineteenth Century*. Adam Hilger, 1986.

Mayhew, Henry. *London Labour and the London Poor*. 4 vols, Griffin Bohn & Co., 1861.

Nead, Lynda. *Victorian Babylon: People, Streets and Images in Nineteenth-Century London*. Yale University Press, 2000.

Owen, David. *The Government of Victorian London, 1855–1889: The Metropolitan Board of Works, the Vestries and the City Corporation*. Harvard University Press, 1982.

Pike, David L. *Subterranean Cities: The World Beneath Paris and London, 1800–1945*. Cornell University Press, 2005.

Porter, Dale H. *The Thames Embankment: Environment, Technology and Society in Victorian London*. Akron, 1998.

Rolt, L. T. C. *Victorian Engineering*. Penguin, 1970.

Saunders, Ann. *The Art and Architecture of London: An Illustrated Guide*. Phaidon, 1988.

Smith, Denis. 'Pumping Stations', *Architectural Review*, 324–328 (1972).

Smith, Denis. *Sir Joseph Bazalgette: Civil Engineering in the Victorian City*. Thomas Telford, 1991.

Smith, Stephen. *Underground London: Travels Beneath the City's Streets*. Abacus, 2004.

Stevens, F. L. *Under London: A Chronicle of London's Underground Life-lines and Relics*. J. M. Dent & Sons, 1939.

Summersom, John. *The London Building World of the 1860s*. Thames & Hudson, 1973.

Trench, Richard and Ellis Hillman. *London Under London: A Subterranean Guide*. John Murray, 1984.

Williams, Rosalind. *Notes on the Underground: An Essay on Technology, Society and the Imagination*. MIT Press, 1990.

Wohl, Anthony. *Endangered Lives: Public Health in Victorian Britain*. Methuen, 1983.

Wright, Lawrence. *Clean and Decent: The Fascinating History of the Bathroom and the WC*. Routledge & Kegan Paul, 1960.

PLACES TO VISIT

As might be expected, London's sewers are not generally accessible to the public and special arrangements usually have to be made to see them. One way is to book a visit during Thames Water's annual Open Sewers Week in May. Website: www.customerhelp.thameswater.co.uk/app/ask

For an insight into sewers outside the capital, organised tours of Brighton's sewers operate regularly between May and September. Website: www.southernwater.co.uk/homeAndLeisure/daysOut/brightonSewerTours

London's sewage pumping stations provide a dramatic architectural insight into the main drainage system, even if they are above rather than below ground:

Deptford Pumping Station, Beck Close, Deptford, London SE13 7RW.
Built from 1859 to 1862, this is the earliest of London's sewage pumping stations and it hasn't survived well: the neoclassical chimney has long since been demolished and the rest of the building is in a sorry state of repair. However, the original buildings used to store coal still survive in good condition next to the Ravensbourne River.

Crossness Pumping Station, The Old Works, Crossness Sewage Treatment Works, Belvedere Road, London, SE2 9AQ.
Telephone: 020 8311 3711.
Website: www.crossness.org.uk/visit.html
The only main drainage pumping station that is regularly open to the public, Crossness survives as a mere shadow of its original splendour, with the Italianate chimney now gone as well as the Mansard roof and clock tower. However, the four enormous steam-engines survive with one now restored and which can be seen 'in steam' on selected Sundays.

Abbey Mills Pumping Station, Abbey Lane, Stratford, London, E15 2RW.
(Tours of the building can be arranged through Thames Water plc, Clearwater Court, Vastern Road, Reading, RG1 8DB.
Telephone 0845 920 0888.
Website: www.customerhelp.thameswater.co.uk/app/ask)
The most lavish of London's sewage pumping stations, Abbey Mills is still an operational building and can only be visited on special pre-booked tours. However, perhaps the best view of the building is from the top of the Greenway – a public cycle and pedestrian route that runs along the top of the embankment that houses the northern outfall sewer.

Western Pumping Station, 124 Grosvenor Road, London, SW1W 8QL.
The most centrally located of the main drainage pumping stations, the Western (completed in 1874) stands prominently, with its original

chimney, on the banks of the Thames near Chelsea Bridge. Still in
operation, the building can only be visited on special occasions such as
during London's annual Open House week in September.

OTHER PLACES TO VISIT

London Metropolitan Archives, 40 Northampton Road, London, EC1R 0HB.
Telephone: 020 7332 3820.
Website: www.lma.gov.uk
The main archives for the Greater London area, this research centre
has by far the most comprehensive collection of records relating to the
planning and construction of the main drainage system, including a
well-stocked open-shelf library.

Kew Bridge Steam Museum, Green Dragon Lane, Brentford, Middlesex,
TW8 0EN.
Telephone: 020 8568 4757.
Website: www.kbsm.org
Housed inside a former water pumping station, this museum has
a large variety of steam engines, many of which have been restored
and are operational.

INDEX